FERTILIZER FOR THE FUNNYBONE

A COLLECTION OF SHORT STORIES AND ANECDOTES

BY
GARY GUEST
(DOUBLE G)

© 2018 by Gary Guest. All rights reserved.

Words Matter Publishing
P.O. Box 531
Salem, Il 62881
www.wordsmatterpublishing.com

No part of this publication may be reproduced, stored in a retrieval system, or transmitted in any way by any means—electronic, mechanical, photocopy, recording, or otherwise—without the prior permission of the copyright holder, except as provided by USA copyright law.

ISBN 13: 978-1-947072-87-9
ISBN 10: 1-947072-87-0

Library of Congress Catalog Card Number: 2018952224

"Sometimes in the ag world today, the ability to laugh at ourselves is the only thing that keeps us from crying. I hope that something I've written here will add to the total number of laughs you accumulate in your life."

Double G

Table of Contents

Acknowledgment vii
Me, A Writer? That Might Be A Stretch 1
Good For A Laugh 5
Sassafras Hill 9
Gassed Out 13
Name Calling 17
What's In Your Gullet? 21
Forget Me Not 23
You Might Be A Cattleman 27
Double G Round Up 31
Ho, Ho, Hog 33
Heads Or Tails 37
Me, A Farmer 41
Move Over…I'm Driving 43
Lofty Expectations 47
Planting Time? 53
I'm Game If You Are 55
Farmer Friends Of Mine 59
I Resemble That Remark 61
Gary The Skinny – Dipper 65
Take My Advice! 69
Big Don't Make It Right 77

Acknowledgment

Going through this life can be difficult. However, it can be a whole lot easier if we are able to share it with someone special. It's even better when that someone special is your spouse.

During the time I put together the contents of this little book, my wife, Denise, has been my typist, my editor, my conscience and sometimes my waitress.

Like they say, "Honey, I couldn't have done it without you."

Thank you, Love You,
Gary

HELLO! AND WELCOME TO MY first book. Yes, I know it's just a collection of short stories, but since it has more than ten pages, I'm calling it a book. That's twice as many pages as some of the books that I gave reports on in English class. Of course, I took freshman English three times.

When I started writing these short humor articles, I envisioned becoming a monthly contributor to some farming publication. Turns out most of these periodicals are making even less money than me, and no one wanted a new columnist.

So one day as I stared at the pile of papers strewn across my desk, I decided to lump them all together in book form. I scooped up all the papers I could carry, went to town and dumped them on my attorney's desk. I could tell I really touched his heart by the way tears were welling up in his eyes. I told him to make me a book. Prior to that day, I hadn't realized that my attorney had a drinking problem. The good news is that since my book is done, he no longer drinks. What a coincidence!

Since cleaning off my desk that day, I've discovered that I'm missing at least two grocery receipts, a service manual for my lawnmower, my 2013 tax return and a complete list of all the girls I've ever dated. Okay, so it was only two names. But it is still a list. And don't listen to my blabber-mouthed friends. I did not make up those names.

So as you can see, it's hard to tell what you might find adorning the pages of this book.

Ever since I was young, I've loved the idea of making people laugh. There are so many things in this world that we can be sad or serious about. Sometimes a little laughter can help us forget about the rest of the world. At least for a little bit.

And so it is that I've compiled this menagerie that you see in front of you right now. While most of the stories are farm-related, some are not. But I hope you will find something entertaining on each printed piece.

If a story or poem makes you laugh, smile or makes you warm inside, then I'll consider myself to be very well-paid. That or you may have gas.

I very sincerely thank each and every one of you and wish you happy reading.

Gary Guest
(Double G)

P.S. If that tax return happens to show up in this book and you read the net income line, I'm just letting you know it was a down year.

Editor's Page

About the Author;

"He's nuts!"

Editor

ME, A WRITER? THAT MIGHT BE A STRETCH

RECENTLY I PENNED A LITTLE humor article hoping to induce a laugh or two from its readers. Since then I've been inundated with countless pleas (Ok, it only counts if your numerical knowledge peaks at five) to offer more writings. I decided to get my wife's opinion. I looked her right in the eye and asked, "Honey, should I make a career move toward writing?" She replied, "Did you eat that last piece of pie? I was saving that." Thusly encouraged, I ventured on. I'm glad that William Shakespeare and Mark Twain have passed so they won't feel the peer pressure from my offerings.

I'm a sixty-year-old cattle and grain farmer from Washington County, Illinois. Denise, my wife of forty years, (Did I say forty? I mean the fourth celebration of our tenth wedding anniversary) and I have four children, ten grandkids, and two dogs. I currently have each one of these mortgaged. Just kidding about the dogs.

The people who know me tend to believe I stretch the truth just a little bit. I don't know. They should see me put my pants on in the morning, now that's some stretch!

I wonder if the farm world could survive without stretching. When farmers are gathered around the feed store, it seems the rainfall totals they recall seeing in the gauge keeps going up or down, depending on whether you are wet or dry. I know some guys whose bushels per acre have gone up so high, they can raise the last hundred bushels without pulling their planter out of the shed. And the snow we used to get in the winter, why it was so deep… wait a minute, I've used that one myself.

I hope using a little creativity when telling a story will always be acceptable. If all we did was tell the unadorned truth, I believe that getting a chuckle out of a written article would be hard, though it might be prescribed reading for someone with insomnia. Being a cattle farmer, I've stretched a lot of fence and sometimes the fence would stretch parts of my anatomy as well. After a day of putting up barbed wire, I would look like I came in second place in hand to hand combat with a bobcat.

Many people used to harvest fur-bearing animals, skinning and stretching them. A good job of stretching would make them worth more.

I chose that analogy with my stories as well. If stretching them a little bit gets me a laugh, then I consider myself well-paid. But lately, I'm afraid I'm losing some of my more consistent audience. Telling my grandkids the "facts" of the way things used to be, was always good for a laugh. But the other day after one of my observations, my five-year-old granddaughter seriously looked me in the eye and said, "Mom says you fib a lot." "Well," I replied, "how do you know your Mom's not fibbing?" Her reply, "Cause Grandma agrees with her." Like Rodney Dangerfield said, "What a tough crowd."

My wife says she's stretching things as well these days. Since we've been married, she says she has to stretch her arms a lot further to give me a hug. Can I help it if her arms shrank?

Stretch ya later,
Double G

GOOD FOR A LAUGH

BENT, CRACKED, WARPED, TWISTED, OFF center. Mention these words around most farmers, and they're strapping their shop aprons on, grabbing their cutting torch and welder and heading for the shed to tackle the repairs on the machine with the aforementioned problems.

When I hear these words, I just figure they're talking about me. Ya, I admit at times I've been every one of these things and then some. My sister-in-law sums it up by saying, "He's just not right!" While all true, in my defense, I say I just like to make people laugh.

We all know that there are plenty of things in life to take seriously. When it comes to family, health, occupation, as well as many others, not every time is appropriate for a laugh. Like when you sit down with a new loan officer, it's best not to open with, "knock-knock." I wish someone would have told me that before. Rejected.

But there are also times when we need a smile or a laugh as being too serious leads to ulcers and baldness. Nowadays with all the men that elect to shave their heads, it's hard to tell someone who has stress and someone who made their barber mad.

My wife and I have many friends. Time spent with two other couples always results in an evening of laughter. They say that laughing keeps a person young. If that's true, the six of us figure we're just entering our teenage years now.

We're all aware the world of entertainment has numerous folks that make us laugh. While so many people in the world of comedy can make me chuckle, two of my favorites are Pat McManus and Mel Brooks. Mr. McManus has written humor articles for outdoor magazines for decades, and Mr. Brooks has turned out some of the most hilarious movies that I have seen. Now that you know who some of my idols are, it might help explain the way that I am.

When I'm trying to induce a smile in someone, more times than not I make fun of myself. And based on my life and the way I farm, I figure I have an unending supply of material to use.

I would like to pass along to you a paragraph from one of Mr. McManus' books. He's talking about drumming up humor, and he says it better than I ever could. Mr. McManus writes,

> "Always include a good supply of humor in your survival kit. Use it liberally for smoothing out rough spots. A dash of it improves the flavor of cold, soggy

sandwiches eaten in the rain. It's a big help in climbing mountains, whatever your particular mountain happens to be. It's good for most kinds of pain, although I personally prefer aspirin. I have found humor excellent for defusing panic, calming hysteria, and easing anxiety. Best of all, it's cheap. You can whip up a batch yourself. Simply take a dab of truth, stretch and varnish it, sprinkle with absurdity and a little foolishness, and salt with enough irony to suit your taste. But never point it at anyone you don't intend to hurt. It might go off accidentally.

As a youngster, I once directed some humor at a tough kid, and he slugged me into a ditch full of ice water. I must say, though, that it got a good laugh out of him and improved his mood considerably."

Here's one more thing that might get you a smile. Find the biggest, gaudiest salt or pepper shaker you can find, and in large letters on the side of the container print, "Laughing powder." Add some jelly beans or something else that rattles and place it on your kitchen table. If someone is having a bad day, pick up the shaker and shake it over their plate of food. Most likely they'll look at you like you're crazy. I should know I get that look all the time. Did I mention I'm a little off plumb?

Stretch ya later,
Double G

SASSAFRAS HILL

Now out on the edge of our nice little town
Where the asphalt streets turn into bare ground
Is the start of a rise that the locals call Sassafras Hill
The elevation is two hundred feet or so
And it's not that impressive as mountains do go
But in the spring of the year, it's always good for a thrill

Now there's a road that shoots right straight up the side
And for most of the year, it's a pretty nice ride
But there's something about this hill that you oughta know

Now most of the year it's baked dry red clay
But when the spring thaw starts and the rains come our way
You'd swear ol Sassafras Hill takes on a red glow

Whenever that clay gets all shiny and wet
Then it's the slickest surface you ever saw yet
And the residents know it's time to just stay away
But there's always some crazys who just gotta try
To drive up that hill though Heaven knows why
But the hill will defeat every truck by the end of the day

The four-wheel drive trucks with their big fat mud tires
And the drivers inside who think they'll go higher
Than the last poor sap who thought that he would give it a go
Each one will try, and each one will fail
And when they exit their cabs, they're a little more pale
And the locals all laugh while telling them I told you so
Now one local lad, Leroy by name
Decided to try his hand at this game
And plotted a plan for him to wake up this town
"While the ground is still frozen, I'll park at the top
And then in the spring, when the road turns to slop
I'll walk up that hill and drive my old pick up down"

Now most people heard of the plan that he had
And they all agreed that Leroy's quite mad
But as spectacles go, they thought that this could be good
So they watched as the snow gave way to bare dirt
Then the spring rain clouds came and started to spirt
And Sassafras hill got as slimy as it possibly could

So Leroy decided that now was the time
And he pulled on his boots, and he started to climb
And he yelled to the crowd "now you all get ready to cheer"
Now Leroy was hoping to have him some fun
So he fired up that truck, and it started to run
As he drove to the brink, he said I'll park right here
Now he was just getting ready to give it a try
When he suddenly noticed the trees passing by
His truck was a flying on down off Sassafras Hill
Leroy was frozen, things happened fast
He shot down the hill and went right on past
The entire downtown and finally stopped at the mill

Now Leroy was certain his hair had turned white
That ten-second ride had caused quite a fright
And to his amazement, his truck incurred not a mark
Now he has no desire to try this again
But I heard him exclaim as he told a friend
"I could have gone faster... if I'd taken the thing outta park!"

GASSED OUT

There's a saying that goes, "Fuel me once, shame on you. Fuel me twice, shame on me." Okay, that may not be the exact way the saying goes, but for the sake of this article, we're going to use it.

I wonder if I'm the only farmer that has machinery with nonfunctioning fuel gauges. I know that some of the gauges on the tractors of my brothers, sons, sons-in-law and nephews don't work either. The reason I know this is theirs is the equipment I borrow whenever mine breaks down. Sometimes at family gatherings, I hear rumblings that their machinery used to work until I borrowed it.

What are the odds that their machine would break down at the time I was using it?

Broken fuel gauges come in different styles. There's the ones that show empty all the time and the ones that always show full. We have one tractor with a digital gauge with little bars stacked from top to bottom. The bars are supposed to disappear as the fuel goes down, but now something has happened, and only every other bar lights up. Either the gauge isn't working right, or there are two-inch layers of air stacked between the layers of fuel.

I don't understand how tractors know this, but the closer they are working to the supply tanks the more acres per gallon you get from the fuel. For example, say you're working a field only fifty or a hundred yards from the farm fuel tanks, and you're down to the last gallon of fuel in the tractor. The tractor will never run out. You could work another hundred acres, and the tractor will keep running. Conversely, you can fill your tractor tank to the top, and once you drive five miles back through the woods to work a twenty-acre bottom, the fuel tank will go dry with one round to go. Best I can tell, if you're far enough away in an area with no cell phone signal, it takes about ten gallons per acre to work that ground.

When it comes to running out of gas, some people are luckier than others. Once my brother, driving his truck without a working fuel gauge, pulled into a convenience store and ran in to get a soda and a snack. He hopped back in his truck, put it in reverse and began backing away...just as he ran out of gas. The truck slowly rolled backward and stopped a hose length from the nearest gas pump. Another customer filling his car shook his head and said he had never seen anything like it.

Another time, my son, coming home from college, had

his car start gasping its last breaths just as he approached a four-way stop at the top of a hill. With no other cars at the intersection, he slowly coasted through it and into the quarter-mile downhill slope. At the bottom of the hill, he silently pulled into the filling station.

Those things never happen to me.

One time I had a dream that it was me who landed on the moon with Neil Armstrong. After gathering our rocks, we got back in the landing craft and prepared for takeoff.

Neil: "Okay fire her up."

Me: "Uhh, It won't start."

Neil: "Why not?"

Me: " Uhh, I think we're out of gas."

Neil: "But the gauge shows full."

Me: "Yeah I know, I got the same gauge on my forty-twenty."

On the moon do you realize how hard it is to find a vehicle to siphon fuel from? Then I woke up. I'm sure it was just a dream, but somehow the next morning I couldn't explain my cornflakes tasting like gasoline.

Stretch ya later,
Double G

NAME CALLING

A COMMON PRACTICE IN OUR WORLD today is for people to be called by a nickname. Sometimes the nickname is self-explanatory, like a guy with red hair might be called "Red" or a tall basketball player might be nicknamed "Stretch." Or in my case, the name my friends have placed on me is "The Incredible Bulk." (That's Hulk. Remind me to chastise my proofreader.)

I thought as farmers, ranchers, and outdoorsmen how often we adorn our animals with names, both affectionately and discriptionaly. (Is that a word?) By that I mean naming an animal based on their size, shape, color, or possibly demeanor. Dog names like Spot, Jumbo, PeeWee or

Weiner. Or cows like Bossie, Freckles, Star and Grandma. Somehow the cows most memorable to me are the ones with adjectives permanently attached to them, like "You no good #**#**##, Kicky, **##**. She left a lasting impression on my mind, as well as my body.

A few years ago when our herds were smaller, it was a more common practice to name each animal. Nowadays most animals are identified by numbers. Hey, there's no reason they should have it better than us people. Today you might hear "You no good #**#**##, number 79, **##**. Well, at least the adjectives stayed the same.

Some years ago I bought a bred beef cow at the sale barn. She seemed calm enough. I arrived at the corral one morning to see her standing over her newborn baby. I climbed into the pen to get a closer look. I think this cow is where the government got their idea for heat-seeking missiles. After managing to stay ahead of her for two laps around the lot, I knew I needed an exit point and fast. I wish someone had been watching. I think it would have impressed them to see a fat man hurdle a four-board fence with a foot to spare. Based on the expression on her face, I know I impressed the cow. I think I could also read on her the look of disgust as well as embarrassment, as she looked around the pen to see if the other cows noticed she let one get away.

One of the comedians then was George Goebel. He often referred to his wife as "Spooky Old Alice." I borrowed that name for my cow. After her calf was a few weeks old, she would calm down and become just Alice. But the next year when she calved again, you always had to know where "Spooky" was.

Some years ago, I used to do a lot of coon hunting. I had a neighbor that had a hound named "Ears." Ears was a pretty fair coonhound, but he had one unusual quirk. Every hunt, Ears would want to catch a possum before getting serious about chasing coons.

It's kinda like going shopping with my wife. First, we go buy a pair of shoes, and then we get started shopping. (Gary's note: Since my wife types all my articles, I'm curious to see if she leaves this in.) (Denise's note: I'm leaving it in…but it's not true.) (Gary's note: I think I just saw a typist turn into a tempest!)

Speaking of names, how many of you husbands have a wife that you affectionately refer to as "The Warden"?

Well, I think that about wraps it up for this month. Our kitchen is rich with aroma of pies baking and homemade bread fresh from the oven. Me, I think I'll go to the garden and pick a fresh quart of strawberries. Then I'll grab my cane pole and stroll through the pasture down to the creek to catch a couple of bullheads for supper. "HOLY CATS!" I think I just turned into John Turnipseed.

Stretch ya later,
Double G

WHAT'S IN YOUR GULLET?

There are so many things that you receive by being a farmer. I thought I would list a few of them.

UNENDING SUPPLY OF CHALLENGES AS we work the soil, raise the crops, and feed our livestock.

LOVE OF MOTHER NATURE WHEN she cooperates and gives us timely rain.

CONTINUAL EDUCATION ON MAINTAINING AND repairing the equipment that we rely on.

EXTRA CLOSE RELATIONSHIPS WITH OUR bankers and lending institutions.

REALLY MEANINGFUL DIALOGUE WITH OUR spouses about the virtues of farm living.

SURE NOT FOR FAIL IDEAS about marketing our farm products at a nice profit.

Review this from top to bottom, and I'm sure you'll agree, that you've received these things just as I have.

Double G

FORGET ME NOT

I HATE STARTLING MY WIFE LIKE that, but the other morning at the breakfast table, I totally shocked her when I shared with her this news.

I said, "You know Donna, I think I'm starting to get a little forgetful."

My wife Denise replied, a bit sarcastically I might add, "You don't say!"

She tried to disguise her amazement with this information by calmly munching on her toast and reading her emails.

I feared that I had already ruined her day so I thought it was best not to belabor the subject. But the truth is

I have noticed myself unable to remember things like I used to. Don't get me wrong, I'm not ready to hide my own Easter eggs yet, but there are little things that escape me.

Actually, speaking of Easter eggs, there was a little puzzlement last year. Our family usually attends church on Easter morning, and while the people stand around and talk after church, one of the adults runs home and hides the Easter eggs in the yard. This past year my wife Debbie said I should go hide the eggs. Mixed amongst the several dozens of hard-boiled eggs is a plastic first prize egg. I'm sure I remember placing the plastic egg in the front yard, so I was a little surprised when one of the grandkids yelled they found the prize egg in the backyard. I cast an accusing eye at the dog. The dog looked back, totally shocked, and shrugged his shoulders. Actually, I'm not sure if a dog has shoulders, but he shrugged something then slunk off towards the dog house.

Truth is, maybe I can hide my own Easter eggs.

I recall a short while ago when I was supposed to stop by the pharmacy. When I arrived there, I called my wife. I said, "Hey Doris, I've got good news and bad news. The good news is I'm at the pharmacy. The bad news is I can't remember why." So she reminded me, and I picked up the product. Actually, I don't know why I use the stuff. I've never once had a reaction that lasted four hours…I was talking about allergy relief…. Why?... What'd you think I was talking about?

I wonder if forgetfulness is contagious. Case in point. The other day my wife Dorothy sent me to town to pick up some milk and cheese. Arriving at town, I decided to stop at the feed store first and see what valuable information the boys there might know. It was kind of a cold day out, and the owner, Jim, and three of his customers had started a pinochle game in the back room. After watching a few games, Hank and Bubba said they had to get going. So me and Ed said we had a few minutes, so we sat in.

Along about noon, Jim's wife brought in some ham sandwiches and coffee, and after eating, I said I had to get going. When I got out to the parking lot, Billy asked me if I wanted to go to his place to see a set of triplet calves that were just born. Well, you don't see triplets that often so I said sure and hopped in Billy's old pickup. As we neared Billy's farm, he announced that he just remembered that he promised his wife that he would bring in a load of firewood this afternoon. Could I help him load it? Well, I still wanted to see the calves, so I said I could help. We loaded the truck, hauled the firewood and stacked it next to Billy's house. Billy's wife was so grateful that she invited me in to share some coffee and cookies with them. After finishing the cookies and warming up a bit, Billy took me out to the barn to look at the calves.

The momma cow and her three little Hereford babies were all doing well, and I told Billy that he was a lucky guy.

While hauling me back to town, we decided to take a few of the backroads and see if we could spy any deer coming out for their evening feed. My truck was the only one left in the parking lot, and I fired it up and headed for home. Arriving at the farm, I headed out to the shed to feed my own calves. I filled the dog food bowl then headed for the house. I hung up my hat and coat and walked into the kitchen and my wife Diane said, "Where's the bread and eggs?" See what I mean about forgetfulness being contagious?

So now I'm putting on my coat and cap and heading back to the store. I figure I'll pick up all four items just to be on the safe side. After all, when you have a wife as special as Denise, you don't want to forget something.

Stretch ya later,
Double G

YOU MIGHT BE A CATTLEMAN

I wrote these few lines in the manner of Jeff Foxworthy. Hope you enjoy them.

IF YOUR MOST MEMORABLE POSSESSION from the cattleman's trade show is a new sorting stick…you might be a cattleman

If you sent your wife (son/daughter) back to sign up at the booth for a second sorting stick…

If the pickup truck that pulls your cattle trailer also is your favorite vehicle to drive to church…

If you've ever went to church with the trailer still hooked up…

If you refer to beautiful things as "pretty as a newborn calf"...

If the paint is worn off of the top rail of the fence where you lean on to look at the cow herd...

If you've ever been to a social event and the conversation turns to manure spreaders...

If you find out all the local news at the feed store instead of the newspaper...

If you're confused as to why your daughter is upset with you because you gave your three-year-old grandson a pair of chaps for a Christmas present...

If cattle get out in the neighborhood and you can identify whose they are just by looking at them...

If the first Christmas sales flyers you look at are from Tractor Supply or Rural King...

If your pickup truck seat has butt depressions where both you and your dog sit...

If you've ever played whiffle ball with your grandkids while wearing cowboy boots...

If you walk into your house and your mother-in-law who's visiting says, "What stinks?" And you don't smell anything different...

If you have a calf in the head chute and you approach its rear end while walking sideways...

If you watch your old herd bull, sell at the auction, and you get the urge to buy it back...

If you can remember the day, you turned the bull in with the heifers but can't remember your grandkids birth dates...

If you have grandkids, who would rather go with you on the ATV to check cows rather than play on the computer...

If you are shopping with your wife at the mall and you

hear a group of young girls pass by talking about halters, and you immediately think of the two heifers you have tied up at home…

If you've ever purchased a Christmas present while attending the local cattle auction…

If you and three city friends attend a cattle auction and you are the only one who understands what the auctioneer is saying…

If your wife tells you to "dress up" and your first choice of neckwear is a bolo tie…

If your grandkids go to school and tell their friends that their grandpa has an unending supply of "Frisbees" in the pasture…

If your wife gets a little pale when the grandkids tell her they have been tossing "Frisbees" in the pasture…

If your children and your grandchildren grow up with the utmost respect for this country and their fellow man, then you can thank God that you **ARE** a cattleman.

Double G Round Up

When suppertimes over, my routine is the same,
The couch and remote, are calling my name
But there's not many new shows that I wish to see,
Most of the new stuff is a waste of T.V.
I surf through the channels, my minds on a quest
To find me the shows that I like the best
At this stage of my life what brings me some joys
Is to click on a station with my favorite cowboys
I'm not talking football, I need a man and a horse
A lady, a villain, and a hero of course
That hero can change, there are quite a few
Whoever they are, they always come through

There's the Duke, and there's Audie, there's Roy, and there's Gene
There's Randolph and Wyatt so you can see what I mean
They'll rescue the lady throw the bad guy in jail
Then off towards the sunset, the couple will sail
Now that kind of show, just ices my cake
Provided of course, that I stay awake
I guess I admit, that sometimes I snooze
My wife wakes me up, says I just missed the news
So often I do miss, the end of my shows,
But I've seen it before, I know how it goes
Yeah oft in my mind, I'm out ridin' the range
But if I could, there's one thing I'd change
While keepin' the law, out west in Dodge City
If I'd a been Matt, I'd a married Miss Kitty

HO, HO, HOG

Is Santa Claus alive and well at your Christmas? I hope so. With four kids and ten grandkids, Christmas time at our house finds Santa Claus's ratings to be right up there with Grandma and Grandpa. I won't tell you who finished one, two and three.

With all the hectic activities of the season, my wife says that controlling the grandkids isn't all that bad, controlling her sixty-year-old child is the hard part.... I have no idea what she's talking about.

I figure that Santa is really just a big old softy. If I had actually gotten all the lumps of coal that I was threatened with, I'm pretty sure I could single-handedly supply the local power plant with coal for a year.

It saddens me nowadays when I read about a "modern" family that doesn't want to disillusion their children with the concept of Santa Claus. In a family like that, I wonder who it is that is actually disillusioned.

Personally, I will never get tired of seeing a child's face light up when they open a special gift.

Around our house, this time of year, the Christmas album that plays the most often is "A Statler Brothers Christmas." I love all their music, but one of my favorite selections is entitled "I Believe In Santa's Cause." And you know what? So do I.

I'd like to pass along a story that I heard about a family living back in the hills a number of years ago. The family made their living with a small number of cultivated acres, some cows, hogs and chickens. Their social status pretty much mirrored that of their neighbors and life was good on that little farm.

Their city cousins considered them to be poor, but as we all know, an absence of cash doesn't mean you're not rich in other ways.

The family consisted of Dad Tom, Mom Betty as well as Sam, age 9 and his sister Sara, age 7.

On this particular year, as Christmas time neared, Tom was using his same old threats about how Santa wasn't coming this year because he was too broke. The kids brushed off this news as they anxiously awaited that special day. One evening over supper, Tom said he had heard Santa was so broke he had sold the reindeer. Even that idea was swept away by Sam and Sara.

Of course, Tom's threats were just that, threats. He had the kids Christmas presents stored in the barn awaiting Christmas Eve. Because the kids were getting a little older,

Tom decided he would do something special this year. He wanted the kids to get a glimpse of Santa dropping off the toy bag on the front porch.

After supper on Christmas Eve, Tom announced he was going out to check the livestock. When he arrived at the barn, he began his transformation into Saint Nick. Tom was a large man with thick black hair and big black beard. Out at the barn, Tom stripped off his shirt and pants revealing his cleanest pair of red long johns. Tom had also snuck out of the house a large bag of flour. He liberally poured it over his head till his hair and beard turned white. He grabbed the bag of presents and started to leave the barn and head for the house.

That's when things went awry.

Gertie was the name of the family's biggest sow, and she had been soundly sleeping in the barn this whole time. Something caused Gertie's eyes to pop open only to discover a strange man with white hair and red clothes sneaking around the barn. Gertie decided to use the barn's main exit at the same time Tom was slipping out the door. Before Tom could say a word, Gertie shot between his legs and headed for the woods. Tom was now straddling five hundred pounds of frightened porker.

Betty, Sam, and Sara were sitting at the kitchen table when they heard all the noise coming from the direction of the barn. They burst through the door and out onto the porch just as the white-haired, red-clad figure shot past yelling, "Whoa, Whoa, Whoa." Of course, the kids heard, "HO, HO, HO."

'Santa' dropped his bag as he passed the house on his way to the woods.

The trio on the porch stared in silence as Santa and Gertie shot into the trees. Sam finally turned to Sara and said, "You do what you want, but I'm sending Santa all the money in my piggy bank to try and help him buy the reindeer back for next year."

We all hope you have lots of 'Dears' around you this season.

From here in Southern Illinois, I'm hoping that each and every one of you has a Christmas that's "High on the hog."

                                              Stretch ya later,
                                                  Double G

HEADS OR TAILS

OK. THAT'S IT! I'VE HAD enough. I think it's time I gave each of you a piece of my mind.... There.... What piece did you get? Oh, that's a good one. That's the part of the brain used for making decisions. I haven't used that since I got married.

Have any of you husbands ever heard this? You and your spouse are deciding what to do, and you give your opinion. Your wife sighs and says, "Oh…well…ok. I guess it's up to you." What this means is it's **NOT** up to you. I decided years ago to reply, "Whatever you want hon." In forty-one years that has been one of my wiser decisions.

That part of my brain that make decisions is seldom

called on anymore because I have a wife, four kids, ten grandkids, two dogs, the FSA office, my FS salesman, my banker, the IRS, the EPA, my church, mother nature and two hitch-hikers who stopped by the farm the other day. I've discovered there are few decisions left to make. Actually, the hitch-hikers opinion that I should paint a giant flower on the side of the machine shed really did add a lot of color to the place.

My wife asked the other day what my plans were after lunch. I said I would probably check the cows, and then head to town to stop by the feed store. She said she had a hair appointment and would really appreciate it if I could wash up the dishes. I said I would. After a little bit, I said to her, "You know what? I don't think I'll wash the dishes right now." Of course, by then she was in the car and a quarter-mile down the road, so she didn't hear me. So I decided to dry them as well.

It was a good decision. I've also decided that I am an extremely lucky husband to have a wife who lets me poke a little fun at her and she still types my stories.

The other night she and I were invited to go to some boring social function in town. While we were changing she seductively snuggled up to me and said, "Do you really want to go to that old meeting or would you just wanna stay home and…well…you know." Well, I decided right then and there that we were staying home. She liked my decision and said she was going to slip into something more comfortable. Well, I'll tell you, I don't doubt that her full-length, one-piece footy pajamas, with the front zipper, are more comfortable. But I also made another decision. I have to learn the combination to that stupid padlock at the top of the zipper!

I've made another decision that I think all of you grain

farmers are going to like. I've figured out a way for all of us to make more money, and it's perfectly legal. It's come to my attention, that the grain market deliberately keeps its prices low until after I've sold my crop. I have a plan. From now on, I'll sell all my crops right out of the field. That way the grain markets can go about sky-rocketing up right after harvest. Every other farmer should get two or three dollars more per bushel, for all the grain they sell. In exchange for this selfless act on my part, every farmer owes me ten cents a bushel for everything they sell. We'll all get rich, and the Chicago Board of Trade will never know what hit them. Keep checking your mailboxes. I'll be sending you a contract to sign any day now.

Now, thus far in this article, I've been leading you on as if my decisions were unimportant or irrelevant. Actually, nothing could be further from the truth. You can ask every single farmer in our area and to a man they will tell you, their very existence hinges on the decisions that I make every year.

My cell phone never stops ringing all winter long with calls from other farmers wanting to know what crops I intend to plant. When I tell them, I've decided to plant eighty percent corn this year and twenty percent soybeans they immediately run to their seed dealers and tell them they are planting eighty percent soybeans and twenty percent corn.

Every morning of spring at sun up, our rural country road is lined with pickup trucks with drivers looking through binoculars to see if I'm pulling the planter out of the shed. When they see me drive out with the planter in tow, they all dash home, put their equipment in the shed and begin to sandbag around their house. I think this is a bit extreme and a little childish on their part. Ten inches of rain in an hour doesn't usually require sandbags.

And the phone calls don't stop come fall. They'll ask, "Are you storing or selling?" I might reply, "I'm sellin corn and storing beans."

At that instant, every elevator in the area is inundated with calls to sell every bean off the combine and to store all their corn for the next three years.

So as you can see, the decisions I make literally keep this county afloat. It's a huge burden, but I've decided...I can handle it.

Stretch ya later,
Double G

ME, A FARMER

The harvest is over, another year,
There's a chill in the air, winter is near.
And as I reflect, on the year that we had,
I guess I would say, it wasn't too bad.
Some yields were high, some yields were low,
If you've farmed long enough, then I
suppose that you know.
That each year is different, and we each try our best,
We plant the good seed and let God do the rest.
But that big bumper crop is not always there,
Some years are hard, no rain in the air.
It troubles us some when things don't work out,

Fertilizer For The Funnybone

We're jealous of others, and our hearts have some doubt.
We sometimes forget, that our blessings abound,
Stop and step back, and just look around.
Our family and friends and this life we've been given,
Are gifts from the Master, all blessings from Heaven.
We couldn't begin, to establish their worth,
It's more than the gold, in all of the Earth.
It embarrasses me so when I fail to see,
All the gifts of my life, that've been given to me.
I'll try to do better, each passing day,
And I hope that I'll stop, and remember to pray.
Lord I praise you forever as I steward your land,
And I know I'm not worthy, of a gift that's so grand,
But I'm thankful to have, this burden of toil,
To raise up a crop, in this beautiful soil.
For I'm eternally grateful, that this was your will
And I know you're beside me, each acre I till,
And each day I'll remember, as I put on my hat,
You made me a farmer…
How cool is that?

Double G

MOVE OVER...I'M DRIVING

How would you like to play a game? Really? Well tough. You have to play, or I don't have a story. The game is called two truths and a lie. It's played by a group of people who one by one state three facts about themselves, two of which are true and one being a lie. Since I'm writing the story, I'll go first.
1. I love being a farmer.
2. I love dirt.
3. I really, positively, absolutely love riding in an airplane and I can't possibly wait till I get the

chance to do it again.

Which one do you think is untrue? Well, I'm sorry to be so tricky with you just learning the game, but the fact is number three is a lie. Yeah, I really don't enjoy flying. I adhere to the philosophy that if man was meant to fly he would have wings. Actually, I'm worse than that. Even if I did have wings, I'm pretty sure I would get back in line and try to exchange them for a motorized scooter.

I really like keeping the soles of my shoes attached to dirt or at least the concrete over the dirt. I'm still alright if I'm riding in a vehicle where the tires are touching the road. Quite possibly I am the only person that can develop air sickness when the car hits a high spot in the road, and we are momentarily airborne. In that one or two seconds, it's really hard for the stewardess of the car to get me a barf bag.

Did I mention that I really like being on the ground? I don't remember much about the science classes I took in high school, but I do recall that they were right after dinner and that forty-five-minute nap really helped me get through the rest of the day. On a side note, do you know if you have the right color pen, you can make an **F** look like an **A**?

As I was saying, I'm not much on scientific theorems, but one thing I've learned through life's lessons is that the farther you are away from the center of the earth, the faster gravity works.

Part of the adversity that I have for flying stems from the fact that the men in my family have always been the ones that want to drive any time we go anywhere. A normal drive to a farm auction usually goes like this. While my dad and my brother are wrestling around for the keys to the truck, I take the spare keys, get behind the wheel and lock the doors until they promise to let me drive.

Myself, along with my wife and two other couples travel quite a bit around this country. Apparently, these other five got worse grades in science than I did because the thought of flying doesn't bother them.

So at the threat of bodily harm, I do some flying. Here's something you may not know. Did you realize it's possible for a human being, on the moment of takeoff, to inhale, then not exhale until three hours later when you land? My wife has video, but if the sight of a person's skin turning blueish-green nauseates you, I wouldn't advise you watching.

The other members of our travel group are extremely sympathetic to my fear of flying and often try to soothe me during a flight by saying things like, "Hey scaredy-cat, the wings just fell off!" If I hadn't been holding my breath, I would have really cussed them out.

One time we were at a vacation spot that we had flown to, I'm not sure where it was, it might have been the moon, when one member of our group, I'll call her Sue, suggested that I write a song about me having to fly. So I did just that. I used the melody of the Dixie Chicks song, "Goodbye Earl." If you're familiar with that song, you can sing the melody while you read these verses.

> The rest of the group is planning vacation,
> all the things they'd like to do
> From San Diego to New York City
> to Texas to Kalamazoo.
> With all these places and wide open spaces
> and with only one asking why
> They made a decision their all on a mission,
> it's simple, Gary has to fly.

Fertilizer For The Funnybone

So now we're flying while I'm still crying,
as we're traveling from here to there
With more white knuckles and tight seat buckles
and with gray growing in my hair.
Now we're traveling around at the speed of sound
and I'm trying not to go insane
So I thought it through now I know what to do,
I'll have to learn to drive the plane.
Did you know the FAA tends to frown on passengers going into the cockpit and asking if they can drive?

Stretch ya later,
Double G

LOFTY EXPECTATIONS

OKAY. I'M GOING TO GET nostalgic here. Many of you will remember this subject while some of you younger folks may not know what we're talking about. Haylofts. That's right. I'm old enough to remember when haylofts were used for...are you ready for this? Hay!

Now some of you couples may have used the hayloft to get away from the prying eyes of your family, but this article is simply rated G. Alright maybe PG. But as far as I'm concerned, you don't need permission from your parents to read it.

Haylofts are simply the upper story of an old barn where feed was stored for the animals, usually small square bales of hay. On our family farm operation, there were five different barns where we kept hay at one time or another. That doesn't count the ten or twenty barns of our neighbors where we used to help put hay up.

There were the four or five friends of our family that always helped one another get their hay put up for the year. But in addition, us farm boys would hire out to other neighbors as well. Some farmers paid two cents a bale. So if your crew got one thousand bales in the loft in a day, you made twenty dollars.

One neighbor would hire me and my friend Lyle to help on a regular basis. This was a dairy farm, so we were putting up hay all season. The bales they made on their own farm normally weighed around sixty to seventy pounds apiece. Lyle and I were in our early teens so these bales were manageable and they helped us develop muscles we didn't know we had.

But in addition to his own hay, this dairyman also purchased hay from a hay broker. This hay was bought by the ton, and I'm sure it was just an oversight on the broker's part that each load of bales came with two tons of water stored inside the hay. These wire-tied bales often weighed between one hundred and one hundred twenty pounds apiece. Since that was more than Lyle and I weighed at the time, getting these bales stored away was a little harder. The two truck drivers that delivered the loads each stood about six foot six and weighed about three hundred and fifty pounds. Their job was to put the bales in the elevator while Lyle and I would try to neatly stack the hay in rows. Every time one of these bales would drop off the elevator and hit the loft floor, Lyle and I were sure the very foundation of the barn was being driven into the ground. Neither he nor I could get the bales more than two high. So we would stack the loft two rows high, then the dairyman told us to liberally scatter salt over the not-to-dry bales. We weren't sure if the salt was to help cure the hay or to slow down the barn fire that we were sure was about to happen. After that, we would walk on top of the first two rows and add another two rows high. Salt and repeat. With all the years of hay that went through that barn, it still looks good and healthy. That's more than I can say about Lyle and me. When my ankles creak, my knees wobble, and my backbone pops, I fondly remember those bales. Okay, maybe it's not fondly, but I said this article was PG.

I wonder how many teenagers today would exert that amount of labor to earn twenty dollars. I would like to think that today's teens are smarter than that. However, having today's youth learning that kind of work ethic wouldn't be all bad.

Now I realize this article is about haylofts, but I must include a short bit about the hay itself. Actually not the hay, but the things that held the hay together. It totally amazes me that some farms can operate these days without baling wire and twine strings. I'm fairly sure that a couple of decades ago the repair bills on our farm would have doubled if it wasn't for baling wire holding our machinery together. You could fix just about anything with baling wire. Your tractor motor could blow up with the engine block in three pieces, and if you could get a couple of wraps of baling wire around the motor, it would at least hold together to finish the field and still drive back to the farm. Baling wire was the predecessor to the modern welder.

Before we reached the age of using the hay bales as a source of income, the bales served us well in our play time. In our younger years, the bales made great building blocks for our forts and castles. Kind of like a giant, itchy version of Legos. We would construct our fortress complete with towers, windows and tunnels. Nothing ever attacked us, but we were ready if something did. That's not exactly true. Once I was attacked in my fort. I was in the loft by myself when I heard something rustling in the loose hay. I poked my head into one of the windows just in time to see a mouse hurdling towards my face followed by a barn cat in mid-pounce. Have you ever seen a cat do a one hundred and eighty degree turn in mid-air? Well, the cat got turned around alright, buts its momentum carried it through the fort window onto the only landing point it could find. The top of my head! Thanks to my ball cap the scars on my head aren't too deep.

One time I told my father I was going to drag a couple of one hundred gallon water troughs up to the hayloft to use in building a moat around my castle. My father suggested I leave the water troughs right where they were, but he promised if he ever saw any dragons lurking about the farm he would help build the moat himself as well as help me raise the drawbridge. My castle must have looked pretty formidable because to the best of my knowledge no dragons ever entered the hayloft.

When I reached the BB gun age, the loft had yet another use. In the winter the rafters of the barn were a common roosting place for the local sparrow and starling populations. I launched countless numbers of these little round projectiles at the birds. While I'm sure some of my shots found their mark mostly what I succeeded in doing was putting dents in the rafters and tin roof. Once while bird hunting in my uncle's hayloft I saw a bird fly toward the far end of the barn. In my haste to chase the bird, I leapt from about three bales high intending to land on the loft floor. My uncle had warned me that many of these floorboards were getting rotten and he was right. My feet simultaneously broke through the floor, one foot on each side of a floor joist. I won't go into details about what part of my anatomy suddenly stopped my descent through the floor but that Sunday in church choir I sang with the soprano section.

Most of the haylofts in our neighborhood came equipped with basketball rims nailed somewhere. Sometimes after 4-H meetings, if there was enough hay already fed out of the loft, a hot game of hoops would break out. It never seemed to fail that sometime during the game; a shot attempt would ricochet off the rim and bust the only working light bulb in the loft. The team that was leading when the lights went out was declared the winner.

I know that there are still barn lofts in use today and I think that is neat. But the era of the two-story barn with the hay stored on top is pretty well gone by the wayside. Nowadays the barn lofts that remain are mainly used by cats and an occasional raccoon. It's sad to see these structures fade away but I know in my heart they made a permanent dent in my life...not to mention my body.

Now I told you when this article started that it was strictly the PG version. Those of you wishing to hear about the R rated version of haylofts should send a self-addressed stamped envelope to my wife. She belonged to a different 4-H club.

Stretch ya later,
Double G

PLANTING TIME?

WINTER IS FADING, SPRINGS ABOUT here,
Open the shed doors, fire up the John Deere,
Crank up the Allis, the Ford and the Case,
It seems planting time, can be quite a race,
Working the land, and planting the seeds
Is the right kind of tonic, this old farmer needs
To shake off the doldrums, of the winter and snow
To put some quick in your step, cause it's bout to go
You grease the old planter, and you tighten a chain
Boy those monitor wires, can sure be a pain
Drive the truck to the shed and load up the seed
We'll put on more bags, than we really should need

Fertilizer For The Funnybone

The tractor's fueled up, and my son changed the oil
Tomorrow we'll put some seed in the soil
We farmers all dream, of the crop that we'll raise
And with all the fingers crossed we'll count down the days
In our minds we're a wishin', for a big bumper crop
We're hopin' the bins get filled to the top
Tonight's sleep will be restless, as I lay there and worry
Did I forget anything, I wish that sunrise would hurry
I throw on my clothes, I just have to go
I step out the door….. into two feet of snow
Mother natures a laughing, as she claps in her glee
Well she got me again, signed, Double G.

I'M GAME IF YOU ARE

I'M SURE YOU'VE HEARD THE expression the game of life. Of course, it's referring to a person's overall time here on earth. In the game of life, like sporting events, you'll have successes and failures and highs and lows. Sometimes your game may take you in a whole different direction than what you originally planned. In today's world pretty much all of us are exposed to different kinds of sports. A lot of these sporting events have announcers. I, myself, broadcast high school basketball for our local radio station. My boss says I have a face for radio.

Wouldn't be funny if we found out that the things we're doing every day were being broadcast like a sporting event? Use your imagination if you will and picture two guys sitting in a booth up in heaven getting ready to call a day in the life of Gary Guest.

"And now, with the call of today's action, here's…John and Bob!" "Thank you, Ed McMahon and good morning to all of you listeners out there! Gary's not awake yet but based on the number of beverages he had last night, he can't possibly lay there much longer. There he goes, like a shot, right out of bed and into the bathroom. We'll use this opportunity to hear from some of our sponsors." "Welcome back. Well, Bob, it looks like Gary's hoping to accomplish a lot on this beautiful springtime day." "You said it, John. He didn't even have a normal breakfast, just eating some leftovers out of the fridge. I'm not sure what kind of effect the pork and beans and sauerkraut are going to have on him later, but on the brighter side, since his tractors air conditioning is broke, he should have the windows open." "Good call, Bob. Well here comes the tractor and planter out of the shed! What a great day to be in the field."

(2 hours later)

"Boy Bob, these rain showers around here really do pop up in a hurry." "That's right John. But on the positive side, Gary can use this time to repair some things in the shop. I've always appreciated watching a craftsman working with a hammer and chisel."

(5 minutes later)

"Well Bob, it looks like the bleeding has pretty well stopped. It's a good thing we weren't recording audio down there, the FCC would have pulled our plug for sure. By the way, did you happen to see how far that hammer flew? " "No, not really John, but it still had a lot of momentum when it went past our booth." Gary's an innovative guy, isn't he?" "That's right John. After finding the band-aid box empty, just like it was the last time he looked, he was able to employ other means of wrapping that finger. I've been told that an oily shop rag secured with electrical tape is a great way to clot up a cut like that."

"While the throbbing of his finger subsides, he can use this opportunity to check the cows. Well, looky there, a brand new baby calf. Say, John, don't you think Gary's getting a little close to that baby?"

"Oh Bob, I'm sure that Gary knows what he's doing. Oh my goodness! Can that cow run! It's a good thing Gary had that five-foot head start, or I doubt that he would still be leading."

"You're right John. Judging by the expression on Gary's face, I think he just jettisoned five pounds."

"I think you're right Bob, but I don't think it will help all that much since the weight is still in his pants."

"As the two racers head for the feed trough, they're neck and neck, and with a dive over the trough, Gary wins!"

"Wow John, that was exciting! And on the bright side, for the moment, I bet Gary has forgotten all about that finger. We'll give Gary a few minutes to catch his breath while we listen to these commercials."

"Welcome back. Well Bob, while we were away, Gary's heart started beating again. Also, he got a phone call from a neighbor telling him it hadn't rained at Gary's other farm. He's going to check it out and see if he can plant there."

"Boy, John, the way that blue smoke is rolling out of Gary's truck, I'd say it's about due for an overhaul."

"Actually Bob, that's not the truck. I think the sauerkraut and beans just kicked in. And if you see Mother Nature walk by, ask her if she can provide enough of a breeze so those fumes don't make it up here."

"Well Bob, it looks like Gary's going to get some planting done today anyway. Even without a guidance system he sure can plant a straight row. Whoa! What happened there? Why do you suppose he suddenly put that figure' S' into his planter swath?" "Well John, I noticed his cab windows were closed. I have a hunch the beans and sauerkraut just resurfaced." "You might be right Bob; it's hard to see straight with tears in your eyes."

"Well Bob, only a couple more rounds and he'll have this field whipped. Uh oh! The tractor stopped. By the way, he's walking to the truck, I'd say the tractor is out of fuel again." " Boy John, you'd think he'd get that fuel gauge fixed. On the plus side, Gary's microphone's still not working."

"Well Bob, the day is coming to a close. It's nice to see Gary and his faithful dog Rover curled up on the couch to enjoy a little TV. I guess it's time for us to call it a day as well. Wait a minute. That boom I just heard. Was that thunder?"

"No John, I think it was the last of the beans and sauerkraut coming out. Hey John, did you ever see a dog gag like that?"

"No Bob. I don't think I have. Well, it was nice to see that Gary's day was so normal. I would have hated to be broadcasting if he had a bad day."

"This is John and Bob signing off."

Stretch ya later,
Double G

FARMER FRIENDS OF MINE

C OME ON IN AND SIT a spell, just grab yourself a chair
John and Gloria sit right here and Duane and Kim right there
These country folks are the nicest ones that you will ever find
Just let me introduce you to some farmer friends of mine
As this big ol world just rolls along and we're livin day to day
It's great to think of all the friends we've met along the way

Now I might be a bit prejudice but maybe you'll agree
That folks who grow up on the farm seem mighty nice to me
Now don't get me wrong there's city folk that can be very nice
Rumor has it that city folks been friendly once or twice
But when your home's on down the road from that city limit sign
There's a darn good chance that you could be a farmer friend of mine
You don't have to own a tractor to be country folk at heart
Just love the land and your fellow man, and you've got a dern good start
Say please and thanks and howdy all and I think that you will see
That you can be a farmer friend to someone just like me
Now if this whole world could get along, well wouldn't that be grand
Let's be the best friends we can be and offer up a hand
If the Lord should ask, "Whom shall I send my Father work to do?"
Well wouldn't it be special if it's a farmer just like you?

I RESEMBLE THAT REMARK

Have you ever seen a television show or a magazine article where they show people that resemble their pets? You know, like a short, stocky guy with a round face and bowed out legs leading his bulldog. Or a tall, slender lady with shoulder-length hair followed by her Afghan hound.

People who study that kind of thing have several theories. One theory is people just buy pets that really do resemble them. Another theory is that over time the pet, and its owner gradually inherit traits from the other.

I decided to put this second theory to the test. Since I'm kind of a large, slow-footed person, I decided to buy a fast and sleek greyhound for a pet. I thought this could be an easy way to improve myself. I can tell you this theory works. I now have a fat, slow greyhound.

It's calving season at our farm, so all my available time is spent in the barn or pasture. My wife told me one day she thought I was becoming a cow. So I decided to jot down a few things to show my wife how a man and a cow are different.

a. A cow is a big, fat, hairy creature…okay, eh, scrap that thought. But….
b. All a cow wants to do is eat and sleep all day…okay, scrap that one, too.
c. A cow is totally indiscriminate as to where to goes to the bathroom and…never mind.
d. A cow always believes that what is on the other side of the fence is better than what's on their own side… Hmmm… this is harder than I thought.
e. A cow's daily agenda is dictated by the demands of its stomach…well, that one was no help.
f. A cow gets all upset just because the doctor is coming, and…okay, well, that needle was pretty big.
g. A cow would rather just shed its hair than get a haircut…well, I am really busy.
h. When a cow goes to its favorite watering hole to quench its thirst, it usually ends up drinking more than it should, and…okay, erase that one.
i. When a cow is hanging out with its friends, it doesn't like the boss showing up and telling it…well, we're just talking.
j. If you're downwind, you can always tell when a cow is approaching…all right, I'll shower.
k. On a hot, sweaty day, a cow doesn't seem to mind the flies following it…I said I would shower!

l. A cow has the ability to get fresh bedding dirty on the very first night, and…well, at least I stayed on my side of the bed.

m. A cow believes that taking a nap under a shade tree in the pasture is a great way to invigorate itself for the rest of the day…I do have really smart cows.

n. A cow might use a tree to scratch an itch it can't otherwise reach…boy, that really does feel good!

o. After a cow rises from a period of rest, flatulence is a common occurrence…oh, man, I raised a stink with that one.

p. When a cow seems to be ignoring you, you sometimes have to raise your voice to get its attention…really, Honey, I heard you!

q. A cow doesn't have enough sense to come in out of the rain…but the fish were really biting just now!

r. A cow will sometimes follow the herd leader into places where they shouldn't be…but the officer who brought me home was very nice!

s. A cow munching on its food with its mouth open is gross to look at and…okay, I'll face the other way.

t. A cow wants to be amorous with its mate only one time a year…finally, something YOU have in common, Honey!

u. Well, it seems my wife might be right! Moooove over Bossy! I guess I'm sleeping in the barn!

Stretch ya later!
Double G

GARY THE SKINNY-DIPPER

(I really like this title, because it's the only time Gary and skinny appear in the same sentence)

As I sit here and think of the years that's gone past,
I can honestly say that I've had quite a blast.
My mind drifts on back, to me in my teens
in an old ragged shirt, and cut off blue jeans.
The midwestern summers could be quite a bear,
The temperatures high and humid's the air.

Fertilizer For The Funnybone

Any exertion could bring out the sweat,
and the shirt you'd be wearing was bound to be wet.
Unless you were lucky and worked in a spot
where the air was conditioned, and the temps not so hot.
But I was a farm kid and cattle we'd raise
and working outdoors just went with the days.
There was cleaning and feeding, and hauling manure.
There was easier jobs, of that I was sure.
But this kind of life is the one that we chose,
but not everyone would agree I suppose.
And sometimes I'd stop and just take a break
in the shade of that oak tree, down by the lake.
I'd shuck off my shirt, my shorts, and my boots
and take me a swim in my own birthday suit.
That water just felt so relaxing and cool.
I thought that the lake was my own private pool.
But I recall a time while swimming one day
I heard a girl's voice, and it was heading my way!
Before I could reach my clothes on the bank,
the neighbor appeared, and my heart kinda sank.
Now Clementine Jones was a classmate of mine,
and her and her folks lived just down the line.
It turns out that she was just out for a walk
and when she spied me, she started to talk.
"Howdy there neighbor, how's the water today?"
That's when she seen my clothes where they lay.
She broke out into laughter as she pointed at me.
Her expression turned evil as she clapped in her glee.
She said, "I'll just sit here and wait for a while."
"I think your pa's calling." She said with a smile.
To sit there and gloat, to me seemed quite rude
and I had no desire to step out in the nude.
So I felt through the waters, hoping to find

some kind of covering to hide behind.
Happiness came when with my foot I did stub
the submerged remains of someone's washtub.
I held it in front, as I walked her way.
I was really quite mad, just what would I say.
Her eyes grew quite large, and her face turned to red
And my blood just ran cold when these words she said,
"I've got news for you, and I'll bet you can't guess.
You're just like that tub, you're both bottomless!"
"Oops" Double G

TAKE MY ADVICE!

WELL, FOR A LITTLE WHILE now, I've been writing some articles and offering up bits of wisdom on a variety of subjects. I thought that perhaps I should give my readers an opportunity to ask me some questions, and allow me to dole out some of the knowledge I have obtained in forty-plus years of farming. Even though I have never attended college, my friends will be quick to tell you that I have a B.S. degree in almost every topic known to man. So I feel obligated to share my B.S. with you. I get lots of very personal questions sent to me, although most of the letters are addressed "Current Resident." I feel just because some are too shy to call me by my name is no reason for me to withhold advice. So let's get started!

1. Donnie writes:

    Dear Double G,

    I am a young farmer, and I am doing my best to raise crops as fine as my neighbors'. Even though my yields are pretty good, every year when we get together and talk, my yields are the lowest of the group. Can you help me improve my yields?

    Dear Donnie,

    Of all the questions that get asked of me, yours is the problem that's easiest to fix. Next year when your group gathers, make everyone else go first stating their yields. Then you give numbers that are bigger than all of them. Remember, one of the most important axioms of farming is, "The first liar never has a chance!"

    Double G

2. Current writes:

    Dear Double G,

    Would you like to win a million dollars?

    Dear Current,

    Yes.

    Double G

3. Resident writes:

    Dear Double G,

    Would you like to make your wife's dreams come true?

Dear Resident,

My wife and I have already discussed this. Her dreams usually require amounts of money that I don't have or feats of athleticism that I can no longer do. So the answer is no.

Double G

4. Bruce writes:

Dear Double G,

What color tractors do you run on your farm?

Dear Bruce,

I'm not sure. I could probably go out to the shed, take out my pocketknife, and scratch through the rust to see if there's any color left. But I'm much too busy answering questions right now to take that much time. Sorry!

Double G

5. Baxter writes:

Dear Double G,

Do you run cattle on your spread?

Dear Baxter,

Yes, and sometimes the cattle run me.

Double G

6. Ron writes:

Dear Double G,

When you go to the bank to apply for a farm loan, what's the one thing that you show them to help you get approved?

Fertilizer For The Funnybone

Dear Ron,

A picture of your wife and kids dressed in gunny sacks.

            Double G

7. Tom writes:

Dear Double G,

I farm fifty thousand acres by myself! What d'ya think of that?

Dear Tom,

(See answer#1)

            Double G

8. Larry writes:

Dear Double G,

My breeding flock of chickens keeps getting mixed up with my Angus herd. What should I do?

Dear Larry:

It never fails. Every time I offer one of these question-and-answer times, somebody sends me a cock-and-bull story like that. Eat more steak and eggs.

            Double G

9. Andy writes:

Dear Double G,

My family and I live on a small farm on the edge of town. My kids have been driving me crazy to get them a horse. The sound of whinnying sends chills down my spine, but the kids love it. What should I do?

Dear Andy,

Get them a pony with a sore throat. Then you can tell them it's little hoarse. (I think Cap'n Stubby would like this one.)

Double G

10. Charlie writes:

Dear Double G,

For a grain farmer to be successful these days, what is one piece of equipment that is essential?

Dear Charlie,

A brand new, four-door, four-wheel-drive pickup truck, with extra lights on top and a fuel tank on the bed. When you're drinking coffee at the local café each morning, you want to look good.

Double G

11. Jeff writes:

Dear Double G,

I have a problem. Apparently, the grass in my neighbor's pasture tastes better than the grass in my own pasture. Every day, my cows jump over the fence and graze all day on my neighbor's land. In the evening, they jump back over and return to my barn. What should I do?

Dear Jeff,

Okay, now tell me again: What exactly is the problem?

                                            Double G

12. Richard writes:

Dear Double G,

My wife and I married a couple of years ago. We thought it would be fun to be farmers, so we bought a little cattle and grain farm. Now I find myself getting up at dawn, working 15 hours a day, and getting into the house totally exhausted. Any words of wisdom for me?

Dear Richard,

I must have your secrets! How do you get by only working 15 hours a day???

                                            Double G

13. Winston II writes:

Dear Double G,

I graduated first in my class from Harvard with a Ph.D. in Psychology. I treat numerous patients with phobias and other related maladies, but I specialize in people with agricultural issues. I become quite upset when overly ambitious and under-educated individuals try to dispense advice in an ultra-vain, glorious manner. Do you agree?

Dear Winnie,

That depends. What does ultra-vain, glorious mean?

                                            Double G

14. Bob writes:

Dear Double G,

I am fit to be tied!! I want to tell my neighbor off, but I'm not sure what to say. The man is dumb, arrogant, self-centered, a know-it-all, stupid, conceited, overweight, a bad dresser, and a poor speller. What do you think I should tell him?

Dear Bob,

Ask him if he's ever considered becoming a humor writer!!?! He seems to have all the qualifications.

Stretch ya later,

Double G

BIG DON'T MAKE IT RIGHT

Well, if you've read through my book up to this point, I hope you've found something that did indeed make you laugh or smile. That was my intent anyway. For my last entry into this book, I'm going to change gears a little bit. What follows is the words to a song I wrote a few years ago. The song addresses an issue that concerns me, and maybe it concerns some of you as well.

In the world of agriculture today, as the farms grow larger and larger; the family farms grow fewer and fewer. The argument might be made that the bigger farms are more efficient, that might be true. But it comes at the expense of a way of life that I feel this country's going to miss. The saddest thing about it, in my mind, is all the kids who won't get to grow up as farm kids. The life values learned on a family farm, usually serve a person well, regardless of where life may take them.

As you read the words of the song. I realize you can't hear the melody that goes with it, but I still hope the lyrics will find meaning in the hearts of some of you folks. Who knows? Maybe in the near future, this song may be recorded.

Thanks for reading my book, and I wish a blessing on each one of you. I hope you enjoy the words to; "Big Don't Make It Right."

Big Don't Make It Right

Verse: 1
I'ved Loved My Life Though It's been hard
Living off the land
We've plowed the fields and raised the kids
Guess it's kinda like we planned
Well I got this love of the country life
From a carin' mom and dad
They Thanked The Lord Each And Everyday
For Everything They Had
If We Were Poor, That Was News To Us
Cause We, Always Had Enough
We'd Put A Few Bucks Back, When Times Were Good
To Survive When Times Were Tough
And Mom Said That It Was Us Kids
That The Farm Was Most Proud Of
Come Drought Or Flood, Seemed Every Year
They Raised A Crop Of Love

Chorus: 1
It Gets Harder Every Year It Seems
To Try And Make Ends Meet
And Farms Have Grown So Large Round Here
And Neighbors Must Compete
For Each And Every Scrap Of Land
They'll Put Up Quite A Fight
They Say Get Big Or Just Get Out

But Big Don't Make It Right

Verse: 2
Daddy Could've Farmed More Land
And Maybe Built More Worth
But He Always Felt, With His Family There
He Was The Richest Man On Earth
He'd Shut The Tractor Down, And Take A Work Day Off
To Take Us Fishin At The Lake
The Best Memories, Ain't The Things You Buy
It's The Memories That You Make
Now A Days The Farms Have Grown So Wide
And The, Daddy's Workin Late
The Mommy's Got A Job In Town
And The Kids'll Have To Wait
Seems Grandma's Recipes Been Lost
Somewhere Down The Line
Take A Barefoot Kid Add Dirt And Love
And Things Will Turn Out Fine

Chorus: 1
It Gets Harder Every Year It Seems
To Try And Make Ends Meet
And Farms Have Grown So Large Round Here
And Neighbors Must Compete
For Each And Every Scrap Of Land
They'll Put Up Quite A Fight
They Say Get Big Or Just Get Out
But Big Don't Make It Right

Chorus: 2
Well I Hope My Kids Can Stop And Think
And Remember Way Back When

Gary Guest

Neighbors Lived, Just Across The Field
And Most Were Next Of Kin
And As We Listen To The Auctioneer
Another Dream Fades Outta Sight
I Wonder If We'll Ever Learn
That Big Don't Make It Right
No Big Don't Make It Right

www.ingramcontent.com/pod-product-compliance
Lightning Source LLC
Chambersburg PA
CBHW052102110526
44591CB00013B/2323